JV E Beginning Readers
EBR Randolph Joa

Randolph, Joanne. Wheel loaders 9000891678

PowerKids Readers:

Wheel Loaders

Joanne Randolph

The Rosen Publishing Group's
PowerKids Press™
New York

For Ryan, with love

Published in 2002 by The Rosen Publishing Group, Inc.
29 East 21st Street, New York, NY 10010

Copyright © 2002 by The Rosen Publishing Group, Inc.

All rights reserved. No part of this book may be reproduced in any form without permission in writing from the publisher, except by a reviewer.

First Edition

Book Design: Michael Donnellan

Photo Credits: pp. 1, 7, 9, 13, 15, and 19 © SuperStock; pp. 5, 17, 21 © Highway Images/Genat.

Randolph, Joanne.
Wheel loaders / Joanne Randolph.— 1st ed.
p. cm. — (Earth movers)
Includes bibliographical references and index.
ISBN 0-8239-6026-9
1. Loaders (Machines)—Juvenile literature. [1. Loaders (Machines)] I. Title.
TS180.5 .R36 2002
629.225—dc21
00-013016

Manufactured in the United States of America

Contents

0891678

This is a wheel loader.

UTAH-PAC
30

A wheel loader shovels and lifts.

A wheel loader has a bucket with sharp teeth on the front. The sharp teeth can cut into rock.

A wheel loader is very strong. It lifts heavy things.

400

This wheel loader carries rocks and dirt in its bucket.

Special tools can be attached to a wheel loader. This tool is like a big claw. It picks up pipes and logs.

A wheel loader has
big tires. The tires help
the wheel loader roll
over bumps.

L-2
WA320
UTAH-PACIFIC

Some wheel loaders even bend in the middle. This helps them turn in small places.

Wheel loaders are used for lots of jobs. A wheel loader is a busy machine!

HOOD

Words to Know

bucket

pipes

tools

wheel loader

Here are more books to read about wheel loaders:

Construction Trucks
By Betsy Imershein
Little Simon

Mighty Machines: Truck
By Claire Llewellyn
Dorling Kindersley Publishing

To learn more about wheel loaders, check out this Web site:

www.komatsu.com

Index

Word Count: 117
Note to Librarians, Teachers, and Parents

PowerKids Readers are specially designed to help emergent and beginning readers build their skills in reading for information. Simple vocabulary and concepts are paired with photographs of real kids in real-life situations or stunning, detailed images from the natural world around them. Readers will respond to written language by linking meaning with their own everyday experiences and observations. Sentences are short and simple, employing a basic vocabulary of sight words, as well as new words that describe objects or processes that take place in the natural world. Large type, clean design, and photographs corresponding directly to the text all help children to decipher meaning. Features such as a contents page, picture glossary, and index help children get the most out of PowerKids Readers. They also introduce children to the basic elements of a book, which they will encounter in their future reading experiences. Lists of related books and Web sites encourage kids to explore other sources and to continue the process of learning.